Introducing ...

Slinky

Wiggles

Willow

Moon

Bud Junior

Buddy

Rose

Budette

Fizz

Millie

Baby buds

The aphids

Adam

The ants

The beetles

Ringo

John

Paul

George

For Teddy and Eva, my buds in bloom – L.H.

BLOOMSBURY CHILDREN'S BOOKS
Bloomsbury Publishing Plc
50 Bedford Square, London, WC1B 3DP, UK
29 Earlsfort Terrace, Dublin 2, Ireland

BLOOMSBURY, BLOOMSBURY CHILDREN'S BOOKS and the Diana logo
are trademarks of Bloomsbury Publishing Plc

First published in Great Britain 2024 by Bloomsbury Publishing Plc

A catalogue record for this book is available from the British Library

ISBN: PB: 978-1-5266-5870-8; eBook: 978-1-5266-6909-4
2 4 6 8 10 9 7 5 3 1

Printed and bound in China by C&C Offset Printing Co Ltd, Shenzhen, Guangdong

To find out more about our authors and books visit www.bloomsbury.com and sign up for our newsletters

Laura Hambleton

Bud

BLOOMSBURY
CHILDREN'S BOOKS
LONDON OXFORD NEW YORK NEW DELHI SYDNEY

At the end of a **long**, **winding** path at the bottom of a **leafy**, **green** garden sits a **greenhouse**.

In this greenhouse live ...

sprouting herbs,

tiny seeds,

 Carrots  Lettuce  Onions

yummy plants,

scuttling insects and ...

... Bud!

Bell Peppers

Don't forget Buzz!

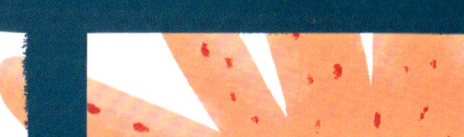

Every morning, Sun **rises**
and **warms** Bud.

Bud's leaves **stretch out** towards
Sun's **nourishing glow.**

How's it
growing, Bud?

Mmmmmm!

Every evening, under the light of Moon, Bud listens to **curious tales** about the **Big Outside**.

Until one day ...

POP!

Bud is planted between **spiky** nettles ...

... and **old** tree roots.

Bud's little roots reach down into the cold, deep soil where worms **wriggle** and slugs **slime**.

Bud is startled by wild weeds, looming trees and unfamiliar bugs. The Big Outside is nothing like Bud had imagined. It's cold, new and scary.

Under the glowing light of Moon, stars twinkle in a dazzling display. Strange sounds surround Bud.

Screeeeeeek

Twit twoooo

Hooooooowieeeee

Squeeeeek

Bud **curls up** below the tall plants
that climb **high** into the night sky.

Everywhere Bud looks, there's a new **danger.**

Big dogs **digging.**

Fast footballs **flying.**

Watch out, Bud!

Hungry birds
pecking.

And naughty insects
nibbling.

Bud misses the cosy red pot in the warm, safe greenhouse. Calling out to the **chattering night**, Bud cries ...

Can I go home now, please?

But **no answer** comes back.

Under the light of Moon's **familiar glow**, Buzz huddles under Bud's leaves and together they fall fast asleep.

With each new day comes a new surprise for Bud.

Pitter.

Patter.

Pattering rain **floods** Bud's roots ...

... and **crashing** storms scare Bud!

CRASH!

BANG!

But look at how Bud grows **taller** and **taller**.

Wild **winds** bend Bud
this way and that!

Whoooosh!

Whoooosh!

But Bud's roots grow stronger
and stronger in the **rich** soil.

Icy **frost** settles on
Bud's growing stem ...

Brrrrrrr!

... but now Bud has enough
leaves to stay warm.

And **Sun** is never far away.

You did it, Bud!

Bud has fully blossomed and bloomed.

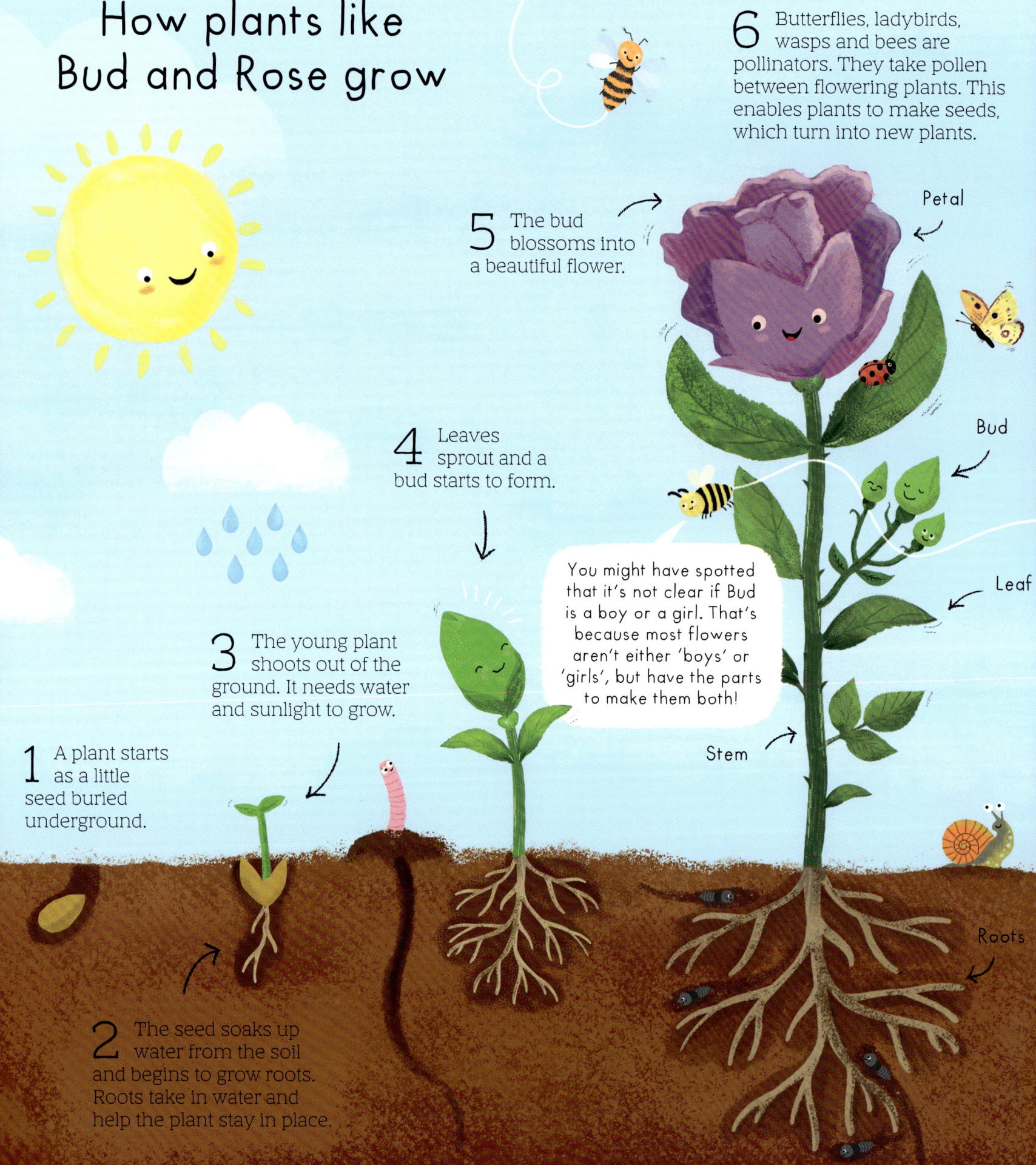

How plants like Bud and Rose grow

6 Butterflies, ladybirds, wasps and bees are pollinators. They take pollen between flowering plants. This enables plants to make seeds, which turn into new plants.

5 The bud blossoms into a beautiful flower.

Petal

Bud

Leaf

4 Leaves sprout and a bud starts to form.

You might have spotted that it's not clear if Bud is a boy or a girl. That's because most flowers aren't either 'boys' or 'girls', but have the parts to make them both!

Stem

3 The young plant shoots out of the ground. It needs water and sunlight to grow.

1 A plant starts as a little seed buried underground.

Roots

2 The seed soaks up water from the soil and begins to grow roots. Roots take in water and help the plant stay in place.

Grow your own plant

Try planting a sunflower and watching it grow!
Sunflowers get very tall and have bright yellow petals.
You will need a small pot or plastic cup, a sunflower seed,
compost or soil, and water.

1 Put some compost or soil in a small pot or plastic cup. Poke your finger in the compost or soil to make a hole. Make sure your pot or cup has a hole in the bottom for water to drain out.

2 Put a sunflower seed in the hole and cover it with the compost or soil. Place the pot or cup by a sunny window.

The best time to plant your seed is between April and May!

3 Water your covered seed every few days or whenever the compost or soil feels dry. Your plant should soon start to grow out of the soil!

4 When your plant grows too big for its pot or cup, ask a grown-up to help you re-plant your sunflower in a bigger pot or outside in a sunny spot.

Measure your sunflower as it grows!